Georg E. Krieger

**Blood Serum Therapy and Antitoxins**

Georg E. Krieger

**Blood Serum Therapy and Antitoxins**

ISBN/EAN: 9783743320376

Manufactured in Europe, USA, Canada, Australia, Japa

Cover: Foto ©berggeist007 / pixelio.de

Manufactured and distributed by brebook publishing software
(www.brebook.com)

Georg E. Krieger

**Blood Serum Therapy and Antitoxins**

# Blood Serum Therapy and Antitoxins.

BY

## GEORG E. KRIEGER, M. D.,

Surgeon to the Chicago Hospital, etc.

*WITH ILLUSTRATIONS.*

Chicago :

E. H. COLEGROVE & CO

1895.

TO THE MEMORY

OF

MY BELOVED FATHER, THE LATE

GEHEIMER MEDICINALRATH DR. E. KRIEGER,

MEMBER OF

THE MEDICINAL COLLEGIUM, ETC., ETC.,

AT BERLIN.

# CONTENTS.

———

# PREFACE.

It has always been the effort of human intelligence to find remedies of specific value in certain diseases and the belief to have discovered a valuable drug has brought upon the market a legion of chemical preparations. Prepared by the chemist, examined by the physiologist as to their effect upon the organism they found their way into the hands of a clinical observer who would soon startle the world by the exhibition of the new " panacea " he discovered. The drug was quickly tried in every hospital, and on the strength of medical testimonials and mercantile enterprise it became in a short time a much demanded "infallible remedy." However as suddenly as it had gained fame, it sunk again into oblivion. This empirical way of detecting specific remedies was well commented on by the following utterance of Rossbach : "Since thousands of years mankind has experimented in this direction and the result has been the discovery of but four remedies for three diseases. It would be a terrible idea that some other thousand years were necessary to detect another four remedies. The usual way of proceeding is too dangerous. Especially harmful proves the enthusiasm not to say dishonesty of many observers. If by chance a physician has found a remedy, after the application of which one or two cases of an infectious disease have quickly recovered, the success is at once attributed to the medicine. No thorough investigation on a larger scale is made, the new "specific" has been discovered and is emphatically recommended.

The result of all this enthusiastic praise is that the profession has lost confidence and can hardly be convinced even if more valuable proofs are offered."

At present we find ourselves in a different situation. We try to explain the etiology of a disease, we endeavor to ascertain the noxious principles and finally to prepare such substances as would act in the opposite direction, and so protect the organism against infection.

The experiment, however, is not made on men before the value of the method has been fully established on animals. Thus having found a preparation which possesses specific curative properties in human diseases, transferred to animals, we try to precisely determine the safety of its application and the limits of its therapeutic value. When after these preliminary studies the remedy is employed in the practice of medicine we need no more investigations as to its specific effect, we know its characteristic influence and only have to corroborate in men the result of former experiments. The only question still open for discussion is the exact determination of the dosage in the various forms of the disease and the different condition of the patients. The above course has been followed in establishing the virtues of serum therapy and antitoxins, the chemical, bacteriological and therapeutic effect of which will be the subject of the following essay.

# CHAPTER I.

## Blood Serum Therapy.

In order to intelligently understand what is meant by blood serum therapy and to comprehend its revolutionizing importance in medical science we have to look back for its origin, and for those facts which are fundamental for this modern method of treating infectious diseases. The principle of serum therapy is based upon the discoveries of Prof. Behring, of Berlin, that the blood serum of an animal organism, which has been rendered *immune* against a certain infectious disease, has the power when injected into another living organism, to protect the latter against, or even to cure it after infection of this disease. For instance, a guinea pig or mouse, which are very susceptible to tetanus poison, can almost with certainty be saved after inoculation with said poison or the virulent tetanus germs if a sufficient quantity of serum of a horse previously rendered immune against tetanus is injected into them.

The practical application of this theory in medicine anticipates a thorough knowledge of the principles of the doctrine of immunity and I, therefore, will consider these in particular before speaking of serum therapy in general.

It is commonly known that one class of animals is more susceptible to a certain disease than another and that some are by nature protected or naturally immune against the same disease. Among those classes that are susceptible for instance to diphtheria, may be some individuals which possess a personal immunity against th e infection. Such personal or individual immunity is either natural or may be acquired. The acquired immunity is effected by accidental or by artificial infection, the former in case of a previous sickness, the latter by protective inoculation. A child who has recovered from scarlet fever is usually thereafter immune. An animal inoculated with anthrax poison is after recovery, rendered artificially immune.

Old as these principles are, one hundred years hav-ing elapsed since they were first applied in medicine by employing vaccination against smallpox (Jenner), they were not understood scientifically until Pasteur made his eminent discoveries concerning the protective inoculation of animals, and by scientific study of the facts thus ob-tained gave the impulse for the development and knowl-edge of a new and highly important science. In 1880, he found that the culture of bacilli of chicken-cholera when dried and exposed to the air for several months, lost some of their virulence, and have not that fatal effect on chickens as when fresh and fully virulent, and further-more that chickens inoculated with such attenuated culture were made immune against the virulent cultures.

At the time of these discoveries it was hardly com-prehended by the profession how important a fact had thereby been brought to our knowledge, but it was soon recognized when the method of protective inoculation showed equally good results in anthrax, erysipelas of hogs and other infective diseases among animals.

Attenuation of cultures.

The principal feature in obtaining the proper material for inoculation is the attenuation of the pathogen cultures. Among the various ways of proceeding, the first was their desiccation by simple exposure to the air, the oxygen of which seems to be instrumental in their attenuation. Soon, however, the latter was more practically accomplished by adding chemical substances to the culture, such as weak solution of sulphuric acid, peroxide of hydrogen, carbolic acid, bichloride of mercury, etc. Further experiments led to another method with as good effect : The attenuation by thermic influence. According to Toussaint (1880), the anthrax culture loses its virulence when kept at fifty-five degrees Cels. for ten minutes.

It took, of course, a long line of experiments to ascertain at what degree the various microbes should be kept in order to render them most useful for protective inoculation, and regarding some of them it has been found that they are best prepared for the purpose by keeping them at a temperature just below the degree which would entirely destroy their vitality. Still another method was obtained by Pasteur's discovery that the microbes of a disease which exclusively belong to a certain class of animals lose their virulence when transplanted to a different species, as for instance those of hog erysipelas to rabbits, those of chicken-cholera to hogs, etc. By all these different methods the living microbes are by no means killed, but their vitality is weakened and their attenuation is accomplished by changing the naturally favorable condition for their development and multiplication for conditions unfaorable to their growth.

The question in which way the inoculation affects the Immunization. organism and how the latter becomes immune has been subject to various theories. At first Pasteur and others were of the opinion that the microbes after having been multiplied, consume the necessary nourishing substances so that a lack of food is caused for those that afterward may enter the system. Another theory and more apt to be correct is that of Chauveau, who believes that during the process of immunization certain bacterial products are formed which are detrimental to the microbes, thereby rendering a second infection impossible. Entirely different, however, became the views on this subject when Salmon and Smith discovered (1887) that immunization can be accomplished without inoculating the living bacteria, namely, by purely chemical agents. This discovery is based upon the fact that the *bacterial products* separated from the microbes had when inoculated the *same immunizing effect* to the organism as a living culture. Thus the chemical nature of the immunizing agent became more apparent, and the change of substance in the organism itself, brought on by inoculation, was considered the real cause for its increased resistance against the respective microbes.

Bactericide power of the Blood.

Simultaneously with this discovery more facts pertaining to the question of protective inoculation were brought to light, partly explaining and partly corroborating the results already obtained. It was observed that the lymph and blood of a normal healthy organism possesses bactericide properties. This fact was substantiated by the experiment to cultivate microbes in the blood of an animal just killed. A large number of the bacterial cells perished and it took several hours before the bactericide power of the fluid decreased to such an extent as to allow the remainder of living bacteria to be developed in the now inactive fluid. Owing to the ingenious investigations of Fodor, Nuttal, Behring, and Buchner, we learned that these bactericide properties are due to certain albuminoids which are dissolved in the blood serum, and furthermore, that the serum of different classes of animals show a different bactericide ability. From these facts the conclusion was drawn that animals which are naturally immune against a certain disease would furnish serum with neutralizing power and the experiments of Behring really proved these to be a fact. Behring found that the blood and blood serum of a rat which is naturally immune against anthrax has strong bactericide properties, while that of rabbits, mice, cattle, etc., which are very susceptible for anthrax infection have none.

The same author in further studying this subject soon Behring's law. obtained results of far more practical value pertaining to immunization. *He established the fact that the blood and blood serum of an individual which has been artificially rendered immune against a certain infectious disease may be transferred into another individual with the effect to render the latter also immune, no matter how susceptible this animal is to the disease in question.* (Behring's law.)

This discovery became fundamental for all the numerous investigations which hereafter were made by many German, Italian and French bacteriologists. The first announcement concerning the artificial immunity against diphtheria and tetanus by blood serum was published by Behring and Kitasato, in the *Deutsch. Medic. Woch.*, *1890*, in which they declared that *the immunity of rabbits and mice when rendered immune against tetanus is based upon the ability of the blood serum to neutralize the toxins produced by the tetanus bacilli.*

Such toxins as will be more elaborately explained under the chapter, "Toxins and Toxalbumins," are the poisonous products of bacterial metabolism, and are the causes for acute disease when circulating in the organism. Their effect is an *intoxication* of the system, while after introduction of virulent germs the cause of the disease is an *infection*. In this case the microbes multiply, and by change of products form a poison which, when absorbed, has a general effect as a secondary result.

In the mentioned article Behring and Kitasato brought experimental evidence for the following important facts : 1. The blood of rabbits immunized against tetanus possesses the ability of destroying the tetanus toxin. 2. This ability remains valid when the blood or the blood serum is taken from the animal.  3.  By transfusion of such blood serum into other animals, the latter and their blood is rendered also immune against tetanus.  4.  The artificial immunity is good not only against inoculation with virulent tetanus bacilli, that is tetanus infection, but also against tetanus toxine, when the latter as a chemical poison in solution is injected. If this was true, the natural expectation was that after the toxin had already poisoned the organism, a sufficient quantity of blood serum of immunized animals would neutralize the circulating toxin, and save the poisoned animal.  This anticipation was thought to be the more correct as experiments had shown that wherever the toxin and immunized blood serum came in contact even in the test tube, the toxicity of the former was neutralized or completely destroyed by the antitoxic properties of the latter.  Therefore, the experiment to inoculate an immune animal with the toxine was now reversed ; that is, an animal not immunized, but suffering from tetanus infection, was inoculated with the blood serum, and the effect was recovery.  Thereby the first evidence was established for the possibility of using the blood serum of immunized animals for therapeutic purposes.

In the following series of experiments it remained to be seen how the serum would effect man in case of infection or intoxication.  It was found that in every instance the above mentioned "Behring's Law" held good ; and further, that the blood serum of individuals, naturally immune against a certain disease, although their blood was resistant against the same, does not possess any immunizing properties for other individuals.  This very important fact proves satisfactorily that the immunizing agent is not a substance produced by nature in those animals, but that the production of it is a result of an organic chemism, to which

the impulse must be given by introduction of the corresponding poison.

What has been said so far regarding tetanus applies
as well to diphtheria and similar infectious diseases. The
diphtheria bacillus also produces toxins, the effect of
which is intoxication of the organism which can be balanced by the properly prepared antitoxin. The latter,
first introduced by Behring and Wernicke, will be considered at length under the chapter " diphtheria."

In accordance with the above mentioned experiments, Immunization
Prof. Ehrlich succeeded in showing that the " Behrings ble poisons.
Law " was valid, not only with infection or intoxication by
bacterial agents but also for some purely chemical poisons,
as ricin, an albuminoid present in the seed of the ricinus
palm possessing extraordinary toxicity and abrin, the toxalbumin of the Jequirity-bean. The animals immunized
against these two poisons by slowly increased doses, furnished blood serum of equally immunizing properties for the
corresponding poison as was shown in tetanus and diphtheria. The gradual increased dose of these poisons rendered the animals more and more immune until a certain
degree of immunity had been reached. That this gradually increasing immunity was not a simple tolerance of the
poison was proved by the fact, that the blood of immunized animals contained an antitoxic substance called
antiricine and antiabrin, which, added to the poison itself,
would attenuate and even neutralize the latter's toxicity.

The results obtained in tetanus and diphtheria, were also
obtained in other diseases, such as different forms of septicæmia and even in hydrophobia, the bacterial cause of
which is as yet unknown. In these particular diseases it
was found as early as 1889, by Babes and Sepp, that the
blood serum of immunized individuals, injected into others
rendered the latter also immune. The immediate effect of

these remarkable discoveries, was a better understanding as to the spontaneous recovery from infectious diseases. It seems that in all of them the blood produces certain substances, antagonistic to the toxins, which had caused the disease, and after the poison had been neutralized the residual antitoxins afford such immunity as is generally observed in scarlet, measles, smallpox, etc. This theory is substantiated by recent reports that blood serum of patients, who had recovered from pneumonia, typhoid and cholera, possessed immunizing power to animals.

uation of serum.

Having thus succeeded in securing a method to prevent, as, under favorable conditions, also to cure persons from infectious diseases, it was of material importance to find :

1st, a mode of estimating the strength of immunizing serum, and,

2d, the best way to procure the material of highest immunizing power. Both problems were satisfactorily solved by Behring, Wernicke, Boehr and Kossel, whose experiments will be given consideration later. In order to secure the highest possible strength, one has to remember that the value of the serum depends on the relative degree of immunity, that is, the difference between the previous susceptibility and the acquired immunity. For this reason it would be proper to take the serum of a very susceptible animal which has been made resistant, but still reacts to inoculation of the poison. The degree of immunity is determined by the figure which indicates how many times the minimum dose, necessary to kill an animal of equal weight, may be multiplied without fatal effect. The immunizing value, however, is given by the figure indicating how much animal weight in grammes can be protected by one gramme of serum, if the inoculation of the toxin has taken place twenty-four hours after the application of the serum. This calculation must necessarily be taken into consideration as it requires a certain length of time before the animal enjoys the full benefit of the immunizing material. In this or similar conditions, which means, as long as the application of serum is made previous to the inoculation, the quantity of serum required to prevent the disease, is much smaller than after the inoculation, and, while in the former case 1 grain of serum of a standard strength called " Normal therapeutic serum " is sufficient, it may require 500 such doses to save an animal but twenty-four hours after the intoxication.

# CHAPTER II.

## Toxins and Toxalbumins.

The above statements may be sufficient to outline the nature and principles of blood serum therapy in general and in the following we will have to deal with the details of this promising discovery of modern medicine. Being confronted in every phase of serum therapy with bacterial poisons the character of which has up to late been unknown, it seems to be proper to pay especial attention to these agents, which are of so material importance in detecting the problem of infection and its cure, before going into particulars of the therapeutic measures.

After it had become an assured fact that tetanus is transferable by inoculation of the fluid discharged from the original wound, Prof. Brieger succeeded (1880) by his ingenious investigations on ptomains and toxins to prove that a certain crystalline substance of great toxicity can be isolated from tetanic fluids and, when injected into animals would reproduce tetanus. The existence of this substance, called tetanin, was obtained not only from animals but also from the arm of an infected patient, a fact which rendered sufficient evidence that such toxins are present in the mentioned disease. This remarkably discovery was immediately followed by Kitasato's equally valuable success in securing pure cultures of the tetanus bacillus.

There was good reason to believe that other microbes produce similar substances and indeed quite a number of toxins were revealed.

The preparation of pure cultures was in itself a warrant for further success in obtaining the chemical material which is essential in transferring infectious disease, and such substances were soon discovered in typhoid, cholera and other pathogen microbes. The effect, obtained by inoculation of the so prepared toxins was not always identical with the symptoms of the genuine disease, neither were the toxins present in every culture not even of the better known microbes. It was therefore natural to search for the factors which were causing this inconsistency of affairs and to ascertain whether the disease depended on still other products. Brieger and Fraenkel, who vigorously investigated these questions, recognized the difficulty which existed, as long as they could not separate the microbes from their toxic products and commenced to experiment with such cultures in which the living microbes were destroyed by heat. But they soon found that with the death of the microbes the whole culture became ineffective and so had to employ another more reliable method. This was found in using a Chamberland or Choalin filter which permitted the liquid substrate of the culture to pass while the microbes of almost every species were kept back by the filter. It was about at this time (1884) when, as a result Loeffler's of many experimental and bacteriological researches, Prof. bacillus. Loeffler found a bacillus which was considered the cause for diphtheria and being an extraordinary good subject for their study, it was this microbe on which Brieger and Fraenkel continued their work in a systematic way. Soon however, they were confronted with another problem. The cultures, usually kept in an incubator for four to six or eight weeks showed when transferred upon another nourishing substrate a different degree of toxicity and sometimes so much less that beyond doubt a change had taken place in these cultures. This gave reason to believe that their toxicity was due to other than such basic substances as repre-

sented by toxins and ptomains.  Both investigators came
to the conclusion, in accordance with Roux and Yersin, of
Paris, who simultaneously worked in the same line, that the
toxic substances produced by the microbes must be of albuminous nature, and after numerous experiments they succeeded in preparing such toxic albuminoids which they
called "toxalbumins." These substances are of different
nature than the toxins and cannot be crystallized, but
represent an amorphic substance.  They were always found
not only in the cultures but also in the diseased organism
and showed at times an exceptionally high degree of
toxicity.

**Nature and preparation of toxalbumins.**     Being without doubt derivatives of the albumen of
tissue they are on account of their relation to the latter of
much more pathological interest.  It was thereby proved
that the metabolism of normal substance can produce material of most detrimental character by means of bacterial
action.  The important influence of the toxalbumins
upon the organism was more positively demonstrated by
the effect of the toxalbumin of anthrax, being the first
prepared as a pure chemical by Brieger and Fraenkel, from
the organs of a diseased rabbit, which had been inoculated
with anthrax bacilli.  The animal, treated with this toxalbumin, showed all symptoms of the original disease.  The
corresponding experiment was made by Dr. Immerwahr
with equal result in tetanus.  The way of proceeding was
as follows: the heart, spleen, kidneys and liver of an animal, just perished after inoculation with Kitasato's pure
tetanus culture, were cut under aseptic precautions into
small pieces and mixed with six ounces of distilled water.
The filtered extract was prepared as advised by Brieger
and Fraenkel until the amorphic substance was obtained.
A solution of the latter, sterilized by filtration through a
Chaolin filter, was in as small quantities as 0.005 grm. in-

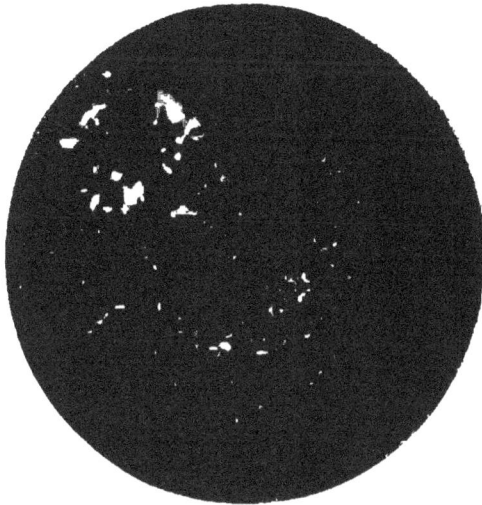

No. 1. Anthrax Bacilli.

The specimen is taken from the spleen juice of a guinea pig infected with anthrax ; stained in gentian violet, magnified 500 times, condensor.

Besides cellular elements it contains the anthrax bacilli, surrounded by a membrane which appears as a capsule with the bacillus as nucleus. The anthrax affection has to be considered a genuine septicæmia, the infection progressing along the course of the circulation. For this reason the bacilli are found principally in the smaller veins and capillary system.

jected into mice which after twenty hours showed the common symptoms of tetanus and died within forty-eight hours.

From the positive result of these experiments Immerwahr concluded, that similar toxalbumins may possibly be obtained even in diseases, the bacterial cause of which has not yet been established. He therefore experimented with the blood, taken from a patient who suffered from a severe form of scarlet fever; however, the effort to secure the chemical preparation, or the inoculation of such blood to mice was of no positive result. In another case of scarlet fever combined with uræmia he could not obtain a substance, charateristic to scarlet but succeeded in preparing an albuminoid which inoculated to animals caused severe spasms and finally death, therefore was considered responsible for the uræmia. Returning to the experiments with toxalbumins of tetanus and diphtheria Brieger as well as Immerwahr and others met with peculiar experience, as they frequently observed a difference in the intensity of the respective toxalbumins.

The same phenomenon had been observed by Roux *Variability of toxicity.* and Yersin with reference to the toxicity of pure bacterial cultures; besides this a series of other questions remained still unanswered : Why is the toxic effect of these toxalbumins so slow compared to that of other strong poisons? In what direction do they act within the diseased organism? What makes the latter immune against an infectious disease after having recovered from same?

During the further study of the subject some important items were noticed which led the experimenters upon the right track in answering such questions. They found that pure cultures in beef juice did not retain their original toxicity for a longer period but became attenuated and at last ineffective even when used in large quantities. In accordance with Brieger and Fraenkel two other investigators —Wassermann and Proskauer—found in 1891 that the cultures of diphtheria bacilli lost their virulence when transplanted from beef juice upon glycerin-agar, that they could however regain part of it as Roux and Yersin demonstrated, when brought back into beef juice.

Discovery of nonpoisonous albumins.

At the same time they discovered in the cultures besides the toxalbumin a second substance of albuminous nature, but different from the former in chemical as well as physiological respect. The new substance was soluble in diluted alcohol and when injected into animals proved to be perfectly harmless. In order to reveal the characteristic properties of both substances it was necessary to separate them from each other, and this was effected by the following process: The cultures in beef juice are filtered through Choalin and the fluid, now free from bacterial cells, is condensed in vacuo to one-tenth of its volume, and for three days kept in the dialyzer with distilled water, so as to separate the peptones and globulins, then again filtered until perfectly clear it is mixed with ten times its volume of slightly sour alcohol (60-70 per cent). The now appearing precipitate remains in the solution for forty-eight hours, which after being filtered is then slowly dropped into absolute alcohol in which a new precipitate appears. Both precipitates are collected and examined separately. The toxalbumin, after having gone through the usual process of dialyzation, etc , appears as a fine white amorphic substance, the other precipitate as a more yellowish brown. The difference between these two substances is most evident with reference to their physiological action. The toxalbumin possesses a high degree of toxicity. The yellow precipitate has none at all. In accordance herewith Proskauer and Wassermann found that very virulent cultures contained a proportionally large

No. 2.  ANTHRAX BACILLI.

Section of the liver of a guinea pig infected with anthrax ;
stained in fuchsin solution, magnified 500 times.

The tissue is not visibly affected, especially not necrotized.  The
microbes are located within the blood vessels.  A small vein and capillary are represented in the specimen filled with bacilli.  In pure cultures the latter form long chains, leaving a small space between two
links which is probably due to a shrinkage, or contraction of the
membrane mentioned above.

quantity of the white substance but little of the yellow, and the ineffective cultures more of the yellow than of the white substance.

The experiments made on animals had the following results: one-eighth of a grain of the white substance, injected into a rabbit, killed the latter in three or four days, one-sixteenth grains in two weeks, one-twentieth in two months. The intensity of toxalbumin, however, was not steady, and showed considerable difference even if the original culture of which it had been prepared had not changed its virulence. For this reason it was impossible to establish the minimum fatal dose. More remarkable, however, is the fact that notwithstanding the extraordinary small quantity necessary for fatal result, the poison has an unusual slow effect. The different physiological actions of the two substances obtained, led to the idea that a certain relation may exist between them, within the infected living organism, and starting from this point of view the above named experimenters tried to isolate them from the organs of animals which had died from diphtheria.

As before, for the preparation of the toxalbumins, the lungs, kidneys, liver and spleen were extracted and submitted to a similar process as mentioned above. These experiments, however, were negative as far as the preparation of the second yellow substance is concerned, and the toxalbumin in this way prepared proved to be still more effective than the one obtained from pure cultures. $\frac{1}{300}$ grm. was sufficient to kill an animal in one to two weeks. The pathological changes observed in such animals by a post mortem examination were correspondent with the usual organic symptoms in diphtheria, except the local inflammation as caused by infection of virulent cultures. Further experiments, intending to show a relation between these two substances and the artificial immunization of animals against diphtheria were at that time also negative, and it took several years before this question could be answered more satisfactorily. The facts so far obtained and proved by experiments were :

1. Infectious diseases are transferred by their specific microbes and the products of the latter.

2. By the action of the microbes toxic substances are produced which are the essential cause for the disease.

3. These toxic substances, called toxalbumins, are of albuminous nature, but have no steady degree of toxicity.

4. The toxalbumins can be prepared as well from cultures from the microbes as from parts of a diseased organism in which they are circulating.

5. The inoculation with either the blood serum of an infected animal or with the toxalbumins brings on similar pathological conditions as in the original case.

6. The existence of another substance, antagonistic to toxalbumin in its physiological effect.

It was especially the last mentioned substance, the true nature of which was then studied with deepest interest, as it was considered instrumental for artificial immunity. Among those who tried to explain this phenomenon Prof. Ehrlich, of Berlin, ranked first. Recognizing the difficulty of experimenting with a pure toxalbumin prepared from microbes or diseased organs, he selected the toxic albuminoids of two vegetable products—ricin, of the seed of the ricinus palm, and abrin, of the jequirity bean, both of which were at his disposal in sufficient quantity and purity. He succeeded in not only immunizing animals against these strong poisons, but also in determining by what method to obtain the highest degree of immunity. In some animals he applied hypodermic injections, others were fed with the toxalbumin, and in some the application of a drop of its solution to the conjunctiva was sufficient for a positive result. He further proved that the immunity obtained by application of gradually increased doses of the toxalbumin is not identical with a mere tolerance toward the poison, and has a certain limit beyond which the resistance against the toxic substance cannot be increased. For instance, a mouse made immune against 200 times the fatal dose died when given an additional dose of the poison, in spite of being continually fed with small quantities.

In the meantime Behring and Kitasato made the fundamental discovery that the immunity in diphtheria and tetanus was based upon a certain ability of the blood to neutralize the toxicity of bacterial poisons, and this led Ehrlich to the idea that the immunity against such toxalbumins as he found in vegetable products was due to a well characterized substance in the blood itself. He at last found that indeed such a substance existed, which he called "antiricin," the injection of which to mice and rabbits rendered the ricin ineffective and the animals immune. It was this most important discovery which became the basis for all the now following researches of the investigators intending to find in the various infectious diseases the analagous substances hereafter called antitoxins. Behring and Ehrlich next established the fact *that the injection of blood taken from immune* animals rendered others also immune. It was a priori evident that the degree of immunity of an animal depended on the amount of antitoxin present in the blood injected. Further experiments intended to determine the quantitative estimation of the antitoxins, and the duration as well as the degree of immunity, opened a wide perspective for this new and rapidly progressing science. Having explained how the improved knowledge of the ætiology of infectious diseases led to more appropriate therapeutic measures than were known heretofore, we will now consider in detail what uses were made of the above-mentioned discoveries.

The two diseases in reference to which the greatest number of experiments were made, and in which the most important results were obtained, are tetanus and diphtheria. The results accomplished is best demonstrated by quoting the experiments of the various investigators.

# CHAPTER III.

## Tetanus.

The first proof that tetanus is an infectious disease was presented by Carle and Rattone in 1884, who inoculated a rabbit with the pus taken from a tetanus wound of a man and thereby reproduced the disease in the rabbit. Soon after Dr. Nicolaier, engaged in examining the microbes of soil, frequently met with the same bacillus, and observed the strange fact that if some of the soil containing the latter were applied hypodermically to guinea pigs and mice, it caused spasms after one to five days ; first locally, later in other groups of muscles. In the pus, discharged from the infected part, he found in every case a long straight bacillus. A similar microbe was seen by Rosenbach in cases of human tetanus, but neither one of the investigators succeeded in finding these bacilli anywhere except in the nearest vicinity of the place of infection, and the experiment to cultivate them from the blood or organs was also negative. The violent effect of these microbes, without increasing much in number nor spreading over the organism, led to the conclusion that they produce a strong poison, and as mentioned above, this was corroborated by Brieger's preparation of a substance from cultures, tetanin, which when inoculated into animals reproduced the disease. He also proved by preparing this tetanus poison from an amputated arm of a patient, that after an infection it circulates in the system. At last Kitasato found a method to prepare pure cultures of the microbe and thereby furnished unquestionable

Tetanus bacillus.

evidence of the bacterial nature of the disease. Further experiments made at the surgical clinic at Halle, in 1891, by Dr. Nissen with the blood serum of a patient suffering of tetanus, showed that mice, injected with very small quantities of such serum, died within a few hours under symptoms identical with those in the original case.

The knowledge of circulation of toxins in the blood which are the products of microbes, caused naturally the desire to nullify their effect by antagonistic substances, *i. e.*, antitoxins, and like the administration of bactericide substances in antisepsis, it now manifested itself a tendency to find the proper antitoxic material. The bactericide substances have on account of this theory not lost any of their importance, remaining as necessary as ever for preventing infection, sterilizing instruments, etc., but in the living organism they are worthless, at least for the majority of cases.

In the administration of antitoxins our aim is not so much to kill the microbes but to paralize the poison produced by them, knowing that, after this has been effected, most any bacterial cell is comparatively harmless as far as the immunized individual is concerned. The cholera bacilli, for instance, which are daily discharged from the intestines of a convalescent are innocuous for him because of his acquired immunity.

Immunity.    With reference to the nature of immunity we have to distinguish two entirely different conditions not only from an ætiological standpoint, but also in regard to their therapeutic value. 1. The protections against certain bacterial poisons may be natural. In this case the natural immunity is caused by the lack of susceptibility to the poisons of such organs as are affected in susceptible individuals. The immunity is therefore hereditary and has no relation to the composition of the blood, consequently it cannot be

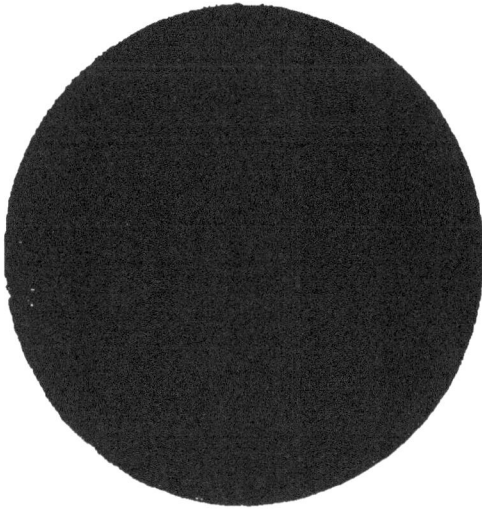

No. 3. Tetanus Bacilli.

Culture on agar ; stained in fuchsin solution, magnified 1,000 times.

The pathogen microbe in tetanus is a slender bacillus with distinct self-motion. Cultivated in higher temperature the germs soon produce spores, appearing on their ends and thereby giving them the shape of a drumstick. Tetanus bacilli are strictly anaërobe microbes, and when transferred upon gelatin, form in the lower part of the substrate characteristic colonies with numerous branches, gradually liquefiying the gelatin.

transferred upon another individual by the blood serum. Such immunity is possessed by white mice against diphtheria, and chickens against tetanus. 2. In contrast to this is the acquired immunity in susceptible individuals accomplished by application of gradually increased doses of a certain poison or by passing through the corresponding sickness, as has been demonstrated in a horse which is naturally very susceptible to tetanus, it being able to stand several million times as much of the tetanus toxin as would kill another horse, not treated before. By such systematic applications of the poison, the blood serum acquires the ability to neutralize its effect, which means ; (a) a chemical change of its composition ; (b) a decreased reaction of living tissue upon tetanus—intoxication. The effect of Effect of tetanus poison to susceptible individuals is therefore a tetanus toxin. threefold one :

(1). The symptoms of a disease peculiar to tetanus.

(2). The production of tetanus antitoxin in the blood.

(3). The modification of susceptibility of the elements to the tetanus toxin.

The symptoms of tetanus are well known and will find their explanation with the last-named effect. The latter, as well as the production of antitoxin has been the subject of deepest study on the part of several investigators, the result of which may be explicated in the following. A remarkable phenomenon, first commented on by Behring, is that the immunity of animals does not de- Immunization of elements o pend upon the presence of antitoxin in the blood but can tissue. remain even if no antitoxin, which gradually disappears from the organism, can be traced any more in the serum. In other words, the resistance against the poison and the amount of antitoxin present, are not corresponding. Some sheep immunized against tetanus, possessed a higher degree of immunity after all antitoxin had disappeared from their blood, than when it contained the largest quantity of antitoxin.   In such a case the condition of an animal can become identical with one that is naturally immune, that is, its immunity can increase to such an extent that an innoculation with even the most virulent culture has no effect, while the blood is free from antitoxins.   Under these circumstances, however, the animal is unfit to have its immunity transferred upon others by inoculation with the serum. It was hereby shown that the presence of antitoxin alone does not only explain the immunity obtained by application of the poison but that certain elements of the organism, which before were susceptible to the disease, have become insusceptible.

Active and passive immunity.

Different from the condition brought on by inoculation with poison, which is an indirect immunization, is the direct immunization by application of the antitoxin.   Individuals so immunized remain immune only as long as their blood contains antitoxin and with disappearance of the latter they become as susceptible again as they were before.   It is necessary to distinguish these two conditions clearly, the antitoxin immunity (passive immunity, Ehrlich) being based upon a temporary addition of a protecting agent to the blood, and the immunity of the elements being caused by an intracellular transformation to the effect of lasting insusceptibility.  The relation between these two different kinds of immunity has been carefully studied by Behring who, in connection with this, established the fact that during the treatment with toxins, animals may acquire even an increased susceptibility instead of getting immune, and at the same time possess a large amount of antitoxin in their blood.   Before this was known a large quantity of antitoxin, found in the serum, was considered evidence for the immunity of the respective animals.   Hereafter, however, it was shown that notwithstanding the production of antitoxin in the blood, an animal may not only be not immune but more susceptible than before. Speaking on this subject, Behring mentions a horse the blood of which contained in one ccm. enough tetanus antitoxin to immunize other horses against such a dose of tetanus toxin, a small fraction of which would be sufficient to kill the former. The knowledge of this fact is of great practical value for the preparation of therapeutic serum, as it shows that the animals most fit to furnish a good therapeutic serum are not those that have the highest degree of immunity but those in whom the inoculation of toxin is followed by a most marked reaction.

The following report of some cases of human tetanus may demonstrate the results obtained in this disease by means of serum therapy.

The first case of human tetanus treated with antitoxin has been reported by Dr. Gagliardi, of Mollinelli, province of Bologna. The antitoxin used in this case was prepared in the laboratory of Prof. Guido Tizzoni and Cattani, two of the most industrious investigators of this subject. The patient, a man forty-five years of age, accidentally contracted a wound on the small toe of the left foot, while going over a rice field. On the next day, May 12, 1891, he consulted Dr. Gagliardi, as the foot had swollen considerably. The Doctor made an incision and treated the wound with antiseptics. On May 19, it was healed. Four days later the patient had symptoms of trismus, which became very pronounced on May 24. Injections of 5 per cent carbolic acid in the vicinity of the wound were of no effect. On June 3, opisthotonus, and all symptoms increasing. At this time the first dose, 0.25 ccm. of Tizzoni's antitoxin, obtained from a strongly immunized dog, was injected, after which some improvement was observed.   On June 7, after relapse and general tetanic spasms, two more injections were made.   The following day the patient gradually improved and was discharged as cured on July 5.   In this case less than 1 ccm. of antitoxin was sufficient to neuralize the tetanus toxin.

Another case was reported by Dr. Swartz, December, 1891.   The patient, a boy of fifteen years, cut himself on the left forearm, August 20, 1891, while opening a walnut which he had picked up from the ground.   Fourteen days later he noticed contractions of the muscles of the arm, and during the two following days also in both legs and the jaw. Applications of chloral and warm baths were of no effect. On September 18, Prof. Tizzoni injected 0.15 ccm. of dog antitoxin, which he repeated twice on the next day.   Hereafter the patient showed signs of improvement.   The injections were repeated on the three following days with 0.25 ccm. antitoxin at each time.   On September 23 the symptoms of tetanus had disappeared and on October 1 the patient could leave the hospital,

Early in 1892 Dr. E. Pacini reported the third case treated successfully with antitoxin.   A peasant, twenty-one years of age, cut himself with a sickle on the left hand, October 25, 1891.   November 4 he had difficulty in opening the mouth, suffered from neuralgic pains and general weakness.   November 13, trismus, but no tetanus of the limbs. During the next six days the symptoms spread over to other muscles and November 19 almost general tetanus set in.   After two injections of 0.25 ccm. of antitoxin for four consecutive days patient began to improve, but was not quite free from tetanus symptoms until December 15. During all this time the muscles of the limbs had not been affected.

The following case, also resulting in recovery, was re-

ported by Dr. G. Taruffi in *La Riforma Med.*, 1892.  A
peasant, seventy-four years of age, while loading a wagon
with wood, severely lacerated his right hand.  Ten days later
the first symptoms of tetanus appeared and increased rapidly during the next two days.  He at once received an
injection of 0.25 ccm. of antitoxin, which caused slight improvement and intense perspiration.  The injections were
repeated daily in the forenoon and afternoon until he had
received six doses, each time with the same result.  His
urine taken before the first injection and injected into mice
had a fatal effect within twenty-four hours; the urine, however, taken after the third injection was innocuous.  The
blood serum obtained by venesection after the second
dose of antitoxin, was also nonpoisonous.  In addition to
these experiments Prof. Tizzoni found in the lacerated
finger, amputated on the same day, tetanus bacilli which
proved virulent to animals.

Equally favorable was a case reported by Dr. G.
Casali, *La Riforma Med.*, 1892.  A woman of twenty-two
years cut her right foot while walking barefooted.  On the
eighth day following the injury she noticed difficulty in
opening her jaws, and six days later the tetanus symptoms
had reached other muscles.  The following three days
she received two injections of 0.25 ccm. antitoxin daily.
After the second injection the symptoms gradually decreased and just one month after the infection patient had
perfectly recovered.  The fluid discharged from the original wound was examined by Prof. Tizzoni who found, besides other pyogen microbes also virulent tetanus bacilli.

By these favorable results, obtained with tetanus antitoxin, quite a demand was created for the same and Drs.
Tizzoni and Cattani had many occasions to test its value.
Although this was very encouraging, they considered the
clinical study of the subject not yet completed.  For this

reason they have so far not published their experience with reference to clinical results. It may however be stated that the mortality in human tetanus, amounting to about 88 per cent in the two largest statistics of Richter (717 cases) and Forgues-Reclus (2,072 cases) has been reduced to 20 per cent and will probably be further reduced with the improvement of this method. A therapy, which can cause such an decrease in mortality, is certainly wonderful and ought to be cheerfully welcomed, especially those who have occasionally to witness the agony brought on by this disastrous malady.

The novelty as well as the unexpected satisfactory results obtained by this method, justify a more detailed consideration of the following questions, which are essential in securing success :

1. What is the difference of the immunizing value of serum of animals of different species? This to know is important, because the immunizing value of the antitoxin or the durability of its effect may differ materially, on account of the constitution and physiological properties of animals of various classes, and we would have to select the most appropriate class for preparing an antitoxin intended for clinical purposes. The experiments made in reference to this, were to the effect that the immunity is the more certain and lasting the more homogen (*i. e.* of similar species) the serum was. For instance, of three rabbits, inoculated with serum of a horse, dog and rabbit respectively, the one injected with the rabbit serum showed the best result. Although being not fully explained, this phenomenon is of great practical value.

*Preparation of therapeutic serum.* 2. Another important item is the preparation of the antitoxin and the estimation of its immunizing value. The procedure is briefly as follows : The virulent germs are cultivated on a suitable substrate, and after being attenuated, either used in that state, or they are first filtered through a Chamberland filter, and thereby freed from bacterial cells. The fluid is then condensed in vacuo, to about one-third of its original volume, and in certain intervals, such small quantities are injected into a horse, as to cause only a slight affection. Later on, more virulent cultures or their poisonous substrate is injected, and when the animal has once acquired a certain tolerance of the poison, the dose is steadily increased, the result of which will be as Behring has shown, that the immunizing strength of the blood serum increases accordingly. The injections are therefore continued until the highest possible degree of immunizing value has been reached.

In regard hereto it is very important, as mentioned above, to know how susceptible the animal was before the inoculation, as it is not the absolute immunity which determines the immunizing value of the blood but the relative immunity, that is the difference between the degree ac. quired artificially and the degree natural to the individual. The immunizing value will therefore be the higher the more susceptible the animal is to the respective infection. For this reason one prefers the horse and goat, which are eminently susceptible to tetanus and diphtheria to other animals for the purpose of obtaining immunizing and therapeutic serum. The selection of the animal, however, is not the only factor which guarantees the preparation of a satisfactory serum ; one has to be positive that the culture with which it is inoculated, is of such a nature as to allow the best possible effect. *Selection of the animal.*

The conditions necessary for this are : The proper reaction and age of the culture in addition to a natural high virulence of the germs. Without observing the necessary precautions in this respect one is liable to meet with a very annoying occurrence, the decrease of the virulence of the *Virulence of the culture.*

culture, which under such circumstances may lose 90 per cent or more of its toxicity from one day to another. An alteration of the toxicity may also be caused by exposure to sunlight, high temperature and by an abnormal alkalinity of the substrate.

Relative to the diphtheria toxin these factors are much more effective in tetanus poison, the original virulence of which can therefore not be as easily maintained as that of diphtheria toxin. The best way of preserving such fluid is by adding 0.5 per cent carbolic acid, or 1 per cent chloroform with which it is kept well sealed in a cool, dark place. The effect the injections have upon the animal, which is to be immunized, is visible after each dose by increased temperature, decrease in weight and a certain change in the blood, which is manifested by the difference of coagulation. The separation of the serum from the corpuscles is retarded and the former is not liquid as usual, but seems suspended in a network of fibrin. During the period of reaction the immunizing value of the blood and consequently the immunity of the animal is decreased; after the reaction, however, both increase and show a further increase after each injection as long as the animal reacts to the poison. Sometimes the decrease of the immunizing value during a strong reaction is so intense that the serum becomes toxic (tetanus producing) instead of immunizing. According to Brieger and Ehrlich the increase of its value is again followed by a slight decrease and hereafter remains steady. In order to secure a high degree of immunity within the shortest possible time revaccinations should be made while the immunizing value is at its height. The animal however must have recovered from its temporary illness as well for this procedure as for the act of tapping by which the therapeutic serum is obtained, because the toxins remain so long in the blood as

*Effect upon the animal.*

the individual shows symptoms of a reaction. Behring and Knorr advise to tap an immunized animal for the purpose of procuring therapeutic serum not before it has recovered its normal condition in regard to pulse, temperature, weight and serum separation. By systematic revaccinations of a horse with tetanus toxin one can easily obtain an immunizing value of 1 to 10,000,000 that means 1 grm. of such serum would be sufficient to protect 10,000,000 grm. of another animal against the minimum fatal dose of this individual. The experiments of Tizzoni and Cattani have shown that the results in dog and rabbit differ materially from those obtained on a horse, the former being by far not as suitable for furnishing a strong therapeutic serum. The comparatively highest value of this serum was : In a horse, 1 to 100,000,000 ; in a dog, 1 to 1,000,000 ; in a rabbit, 1 to 1,000,000.

We take occasion at this time to emphasize the difference between "immunizing serum" and "therapeutic serum." The former is used for the purpose of protecting an individual against a possible infection and intoxication, or at least to mitigate its effect. Therapeutic serum is applied with the intention to neutralize the toxic effects of an infection or intoxication which is already in progress. It is readily understood that the serum used for this purpose must be many times stronger than immunizing serum, and it has been found that certain proportions exist between the two kinds, so that if the value of the immunizing serum is known one can calculate its value as therapeutic serum and thereby determine the quantity required for clinical purposes. The therapeutic value of the serum is often not more than $\frac{1}{100}$ or $\frac{1}{1000}$ of the immunizing value and decreases according to the time which has elapsed between the tetanus infection and therapeutic application. Kitasato who has studied this question very closely, observed the following facts :

*(margin: Immunizing serum. Therapeutic serum.)*

Ten mice infected with tetanus germs showed, forty-eight hours later, distinct symptoms of tetanus.  At this time each of them received one ccm. of serum of an immunized horse.  Notwithstanding this comparatively large dose five mice died eighty hours after the infection.  The other five were again inoculated with the equal amount twenty-four and forty-eight hours after the first injection. They recovered very slowly but steadily, so that after several months they were in normal condition.  When instead of forty-eight hours the same quantity of serum was injected twenty-four hours after the infection all animals recovered. If only twelve hours had elapsed the tetanus symptoms were very light and could be checked by less than one-half of the dose.  If the infection took place fifteen hours after instead of before the inoculation 0.001 of the same serum was sufficient to protect a healthy mouse against the tetanus, and after simultaneous application of the toxic and antitoxic substance one ccm. of the latter prevented the development of the disease.  It was thereby clearly shown that the sooner the serum is applied the smaller quantity is required to have the same therapeutic effect, and that the amount necessary to neutralize the toxic effect can be determined according to the time which has elapsed since the infection.

It was natural that after ascertaining these facts another question arose which could be answered only by the most minute and delicate investigation, namely : Upon what action within the living organism is the curative effect of the antitoxins based ?  In regard to this question two different opinions prevail.  According to Behring and Kitasato, the bacterial poisons are directly neutralized by the antitoxins.  The fact brought forth as proof for this theory was, that when the poison and the antitoxin are mixed in certain proportions in a test tube the effect of such mixture injected into an animal is usually negative. This observation is apparently corroborated by Fedoroff who found by a series of experiments, that if the tetanus poison and the antitoxin mixed in the proportion of 1 to 2 is injected into rabbits, the latter are but slightly affected ; if, however, the amount of tetanus poison contained in such mixture is injected separately in a different place, but at the same time with the double quantity of antitoxin, the animals usually die. He is therefore inclined to believe that the action of the poison is directly balanced by the antitoxin.

Contrary hereto are the results of Buchner's experiments. Buchner inoculated twenty-three white mice with a certain quantity of tetanus antitoxin, prepared as a dry substance from a beef tea culture, and also with an equal quantity of antitoxin, likewise prepared, to the effect that only three mice died, and of the twenty remaining, eleven were affected slightly and nine not at all. The same experiment made on rabbits, which are much more susceptible to tetanus than mice, had the effect that of twenty-three animals eight died, while the others were but little affected.

From these facts Buchner concluded that the effect of both substances to each other is not a direct one, but is based upon a certain influence to the living cells, which, being affected in two opposite directions, remain neutral. This theory is substantiated by several facts. 1. The long time before the antitoxin acts. 2. The slow disappearance of the tetanus symptoms. One would naturally expect that every case of tetanus, even in an advanced state, could be checked by antitoxin, if it were a purely chemical antidote. This, however, is not the case, as it was observed that the therapeutic serum, even in a mild form of tetanus does not relieve the contractions present, but only protects other parts of the organism against the poison, thereby limiting the action of the latter. This would explain why it has no effect in advanced stages of the disease, and why a relapse can occur in case of incomplete immunization.

The following experiment of Tizzoni corroborated this theory: Some animals were inoculated with sufficient quantity of a virulent tetanus culture to cause their death if not immunized. On the next day, after the first symptoms had appeared, they received antitoxin. In spite of this, the local symptoms increased, but being otherwise protected, the animals recovered. Thus the tetanus poison was limited to its local effect, but not completely destroyed, until eliminated by the organism. For these and other reasons it seems that the living cells take a vital part in producing immunization. Another point which attracted considerable attention was the question, which part of the organism is affected by tenanus virus and antitoxin. Bruchettini found by his researches regarding the distribution of the poison within the animal system, that from the place of infection the toxic substance is first carried away by the blood, but that the larger glands, with the exception of the kidneys, do not absorb the poison, and that part of it is eliminated with the urine.

Other investigators, as Valliard, Vincent, Autocratow, and others showed by experimental evidence that the muscles, even in severe cases, do not contain any toxic substance, and that the latter causes muscular contractions merely by its effect to the nervous elements. To the same conclusions came C. Brunner, whose clinical studies of the subject revealed:

1. That the tetanus poison does not affect directly muscular tissue, the spasms of which can be relieved by curarizing the ends of the motor nerves.

2. That no spasms can be caused by tetanus poison in muscles, the motor nerves of which have been separated from the centers.

3. That the muscles of a certain nerve territory react on the poison only if the corresponding nerve center is irritable.

4. That the spasms are not the effect of irritations of the sensitive nerves, but of the central organs.

With reference to the action of the antitoxin, the results of various investigators showed that the blood is the first but usually only the temporary carrier of the immunizing substance, and Tizzoni's assistant, Dr. Centanni, in his researches regarding the specific immunization of the elements of tissue, came to the conclusion that the latter and not the blood is finally the one which is responsible for the protection of the organism.

Corresponding with the experimental results in regard to the effect of *toxins*, it is the nerve element which seems to be the principle absorbing place also for the antitoxin. It is, therefore, evident that if antitoxin is applied to neutralize the effect of the toxin, both substances will fight for their existence on the same field, and that the ground held by the one cannot be occupied by the other until the former has weakened or is eliminated from the organism.

The last problem to be solved in the doctrine of serum therapy, was regarding the possibility of inheriting an acquired immunity against tetanus. Ehrlich and Huebener have most elaborately investigated this subject by a large series of experiments, the results of which were as follows: If the offsprings of a normal, that is not immunized mother, and an immunized sire were inoculated with the minimum fatal dose of a tetanus toxin, they invariably died within the usual time. The descendants, however, of a mother which was immune at the time of conception, and a normal sire was not affected even by 100-times the fatal dose. This clearly shows that the immunity can be inherited from the mother but not from the sire. The duration of such inherited immunity was found to be comparatively short, lasting usually but two months, and in no case longer than three months. These results are corresponding with those obtained by Ehrlich in his experiments with vegetable toxalbumins, a fact which is considered important for the knowledge of the real nature of toxic substances in general.

In conclusion we wish to recapitulate the principle points to be considered in protecting man against the effects of tetanus infection. The most important factor with which we have to deal is the gravity of the infection and the time since the latter took place. The gravity of an infection is measured by the more or less rapid effect upon the various groups of muscles.

In regard to the treatment of a person recently infected the following rules seem to be advisable :

1. As the reaction is usually proportional to the quantity of poison introduced into the organism, an early and thorough antisepsis should be observed in dressing the original wound.

2. The best means for the efficious neutralization of tetanus germs and toxin have been found to be a strong solution of sublimate with tartaric acid and especially the actual cautery.

3. The amputation of the infected part, even if the latter is of minor importance, will hardly afford any benefit, because the soluble poison is rapidly absorbed and carried into circulation.

4. As soon as the place of infection has healed, no beneficial result can be expected from a local treatment,

and no time should be lost in using the antitoxin in order to effect the best possible immunity.

5.  The quantity of antitoxin to be injected is determined according to the strength of the latter, the gravity of the infection, provided there are visible symptoms, the time since the infection occurred and the age of the patient. The shorter the time of incubation or the more serious the latter, the larger ought to be the first dose of antitoxin.

6.  The number of injections necessary to save the patient depends also on such circumstances as just mentioned, but as a rule one injection should be applied every other day with a somewhat smaller quantity than used at first, in order to maintain a sufficient immunity and to prevent a relapse, which occasionally occurs even after the immunity had apparently set in and which is due to a premature elimination of the antitoxin.

# CHAPTER IV.

## Diphtheria.

Far the greatest interest has been shown by the medical world as well as by the laity to those discoveries and experimental results in blood serum therapy, which pertain to the question of protection against diphtheria.

Before we discuss however this question in particular we deem it proper to reconsider the theory of the double nature of immunity, which is fundamental for the results obtained by blood serum therapy and thereby wish to emphasize the difference between natural and artificial immunity. The theory of natural immunity is based upon the fact that the blood and blood serum posseses certain bactericide properties, as has been lately proven beyond doubt by Denys, Kaisin, Vaughan, Kossel, Hankin, Buchner and others. These bactericide properties are due to soluble substances which are circulating in the blood, called "alexins." The efficacy of such substances, although preëxisting, can be raised to a higher degree by inoculation of germs or their products. Concerning the nature of the alexins we know very little, but it is supposed that like the toxic bacterial products they belong to the albumins. In regard to their origins the experimental investigations of Vaughan, Kossel and Buchner have furnished evidence that they are produced within the leucocytes. This, however, is not identical with the theory of Metchnicoff, who found by direct microscopical observation, that the leucocytes attack the germs, take them into their interior and destroy them by digestion. But such occurrence, called phagocytosis, which has been most elaborately described in Senn's " Principles of Surgery," cannot, as Metchnicoff believed, be an unquestionable proof for his theory, because it is known that not only some kind of germs are not attacked by the leucocytes, but that even if the latter have been frozen to death, the blood serum still possesses its former bactericide properties. It is thereby proved that phagocytosis, wherever it is observed, is a mere secondary feature and that the alexins retain and sometimes even acquire a higher degree of bactericide power by freezing the serum. The power to effectually attack the germ must be attributed to dissolved chemical substances, produced by the white corpuscles. The fact that the latter frequently take the germs into their interior seems to be only an intention of nature to apply such substances most effectively upon the bacterial intruder, and in this way both explanations for the bactericide influence of the blood serum stand to reason.

The experiments made by Dr. Vaughan to determine the nature of the germicide constituents are leading to the conclusion that the active substance is a proteid formed in the nuclei of the cells, and therefore called nuclein. The effect of this substance on cultures of cholera, anthrax and staphylococcus pyogenes aureus proved, according to Vaughan, in every case its germicidal properties. After the fact was established that such bactericide substances are produced in the white corpuscles, another observation heretofore misunderstood is now readily explained—the numerous affluence of leucocytes toward a place of infection, and their abnormal increase in certain infectious diseases. All these phenomena mean nothing else but a defense with which nature provides our organism against the bacterial offender.

The substances so produced are of very unstable character, and must be different in various classes of animals as well as in different individuals of the same species, because they afford, according to their antagonistic power against certain microbes, the protection which is known to be natural to the respective class, and, as mentioned above, called natural immunity. For instance, as rats are naturally immune against diphtheria, the alexins prepared in their leucocytes must differ from those prepared in the white blood corpuscles of chickens, which are susceptible to this disease, but are immune against tetanus, which would seriously affect the ràts.

**Nature of the antitoxins.**
Entirely different from the cause of natural immunity is the process by which artificial immunity is obtained. We have seen in our first chapter that immunity is the result of the production of antitoxins within the organism or their application to the organism. Considering the nature of these antitoxins from a biological standpoint, we have to take notice of their great resistance against such influences as light, heat, and even decomposition, a fact which is usually not found in animal products, and which rather corroborates Buchner's theory that they are of bacterial origin. In accordance herewith is Behring's and Knorr's experience that a tetanus culture, after being attenuated by exposure to 65° C., sometimes shows directly antitoxic properties. Similar observations have been made by Brieger, Kitasato, Klemperer, Wassermann and Nencki.

It further renders the relation between the toxalbumines and antitoxines at least possible, if not probable. The effect of both substances upon each other has been discussed in the chapter on tetanus, and it may only be repeated that very likely the living cell is instrumental as well in producing the antitoxin as in rendering the latter, when introduced as such, effective against the toxins, no matter if they were directly inoculated or originated from an infection.

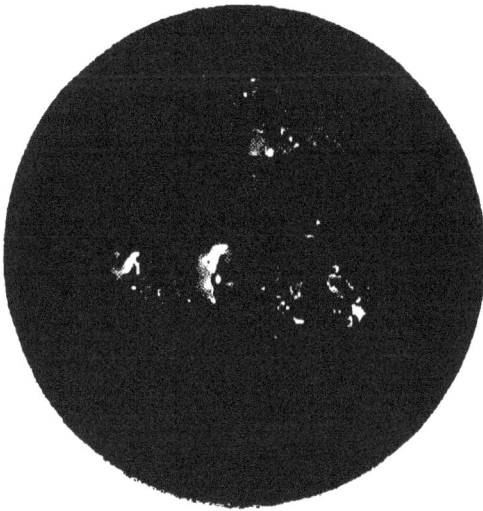

No. 4.  Diphtheria Bacilli (Klebs-Loeffler).

Section of a diphtheria membrane taken from a tonsil ; stained in alkaline, methylen blue, magnified 500 times.  Immersion.

The darker spots represent thick accumulations of *diphtheria bacilli* near the surface of the tonsil.  In the deeper layers the bacilli appear more separated.  The diphtheria bacilli are found exclusively at the point of infection, especially on mucous membranes.  They never appear in the blood or the organs.

The line where the dark spots join a lighter field represents the limit of necrotized tissue (C. Fraenkel).

Besides this, it should be remembered that the antitoxins are limited to a specific effect upon the corresponding poisons, which is not the case with the alexins. The
latter, however, are acting in harmony with them when not
toxic products, but the germs themselves are the object of
offense. With the exception of this occurrence we have
two well characterized conditions—those produced by the
alexins, natural immunity, and those produced by the antitoxins, artificial immunity. The principal difference between these two groups of substances is that the former are
of purely animal nature, while the latter most probably
originate from specific bacterial products. In the immunized animal the antitoxic principle is, however, not limited
to the blood serum or the leucocytes, but is present as well,
and occasionally exclusively, in the tissue cells. It is evident that under certain conditions both peculiarities, natural immunity, which would better be termed natural tolerance or resistance, and artificial immunity, exist and act
in the same organism for the same purpose, and although
each may be raised to a higher value, they will always
retain their distinct character. Natural resistance can
never become equal to specific immunity, nor can specific
immunity become a natural resistance. The chief practical difference between these two conditions, however, is
that the latter is not transferable to other individuals,
while specific immunity can be transferred by the inoculation of the serum from individual to individual.

Next to the knowledge of these facts, the administration of antitoxin demands a certain experience concerning the contagion, intoxication and virulence of diphtheria bacilli and their poisons. As generally known, the
germs which are now called diphtheria bacilli were discovered by Klebs and Loeffler in 1884, and soon hereafter
the experiments of Roux and Yersin corroborated the ob-

servations of the first named author, that wherever genuine diphtheria exists these germs are present, and that when inoculated from one animal to another they reproduce the disease.

While investigating these important facts, Hoffmann found in 1887 in many instances besides the genuine diphtheria bacillus a similar one which exactly answered the description of the Loeffler-Klebs bacillus, but which after being isolated and inoculated to animals proved perfectly harmless. He therefore called the latter pseudo-diphtheria bacillus. In regard to the question whether this germ is different from Loeffler's or only an attenuated genuine diphtheria bacillus much has been written, and the majority of bacteriologists are inclined to approve the opinion of Roux, Yersin and C. Fraenkel, who believe that though a marked difference is noticeable between the two bacilli as well with reference to their virulence as occasionally also in their form, these differences are not sufficient to characterize the bacilli as two different species, but only as varieties of the same kind. This opinion seems the more plausible if one considers the largely fluctuating virulence found in genuine diphtheria bacilli, which is according to Roux-Yersin and Brieger-Fraenkel the most important cause for the varying gravity so often observed in cases of diphtheria. As the space forbids to go into details about the etiology of this disease we wish to mention only one item which is of importance in explaining not only the varying gravity of diphtheria infection, but what is of greater interest to us, the reason why in some cases the antitoxin will fail to have the desired effect. This occurs when an infection of other malign germs exists simultaneously with that of the Loeffler bacilli.

Mixed infection    It is an acknowledged fact that the virulence of diphtheria bacilli increases in the presence of streptococci.

No. 5.　STREPTOCOCCUS PYOGENS.

Streptococci and leucocytes of human pus ; stained in gentian violet, magnified 1,000 times (Pfeiffer and C. Fraenkel).

The streptococcus is a globe shaped chain forming microbe.　It is
found in almost every suppurative process, and is the cause for various
pathological conditions.　Introduced into the subcutaneous tissue and
the lymphatic system the streptococci cause the characteristic affection
known as erysipelas (Fehleisen).　In diseases of the inner organs the
microbe takes a prominent part on account of its virulence.　The
streptococci frequently accompanies diphtheria bacilli, being an essential factor as to the malignancy of the case.

This being known the question was whether the increased virulence is due to a stimulation of Loeffler's bacilli by the streptococci to produce more toxin, or if the organism is rendered more susceptible to diphtheria by the other germs. In order to answer this problem Funk, who has devoted considerable interest to the study of such mixed infection, made two series of experiments in the Berlin institute of infectious diseases.

First he tried to ascertain what effect a certain quantity of streptococci would have when inoculated to a guinea pig with diphtheria toxin of a tested strength. The effect was measured by a quantity of antitoxin sufficient to neutralize the toxin and the experiment led to the result that the animals did not show an increased susceptibility for the toxin. Of more positive effect, however, were the experiments made with streptococci in combination with living diphtheria cultures instead of the ready toxin. The antitoxin used in the animals, inoculated with this mixed material was in no case sufficient to protect the animals against infection, which resulted in death in about 50 per cent, while the animals which received an equal amount of diphtheria cultures and antitoxin but no streptococci all remained well. Thereby the interesting fact was established that the streptococci have a stimulating effect upon the productive ability of diphtheria bacilli, which, however, can occasionally be compensated by an accordingly increased dose of therapeutic serum. We shall return to these important facts when speaking of the way in which a neutralization of the diphtheria poison is effected by the antoxin. Being familiar with the phenomenon that conditions as mentioned above result in a different intensity of diphtheria infection, a practical use of this fact was first made by Behring and Wernicke, and later by H. Aronson who endeavored to ascertain the degree of virulence of different cases and succeeded in determining how to obtain pure cultures of a standard efficacy.  Aronson further found, that the cultures from cases of different severity were of so much varying virulence that in one instance 1 to 2 grms., in others but 0.05 grm., of a culture was necessary to cause death to a guinea pig.  These results of experimental studies explain very clearly the variety of single cases as well as of whole diphtheria epidemies.

A material progress in the researches regarding the virulence of diphtheria bacilli was made by Roux and Yersin's observations that the virulence can be raised by cultivating the bacilli either on a suitable nutrient substrate together with streptococci, or by passing the germs through a series of animal organisms, a method by which under favorable conditions as to the individual susceptibility, etc., a virulence 100 times stronger than that of the original culture can be secured.  The contrary effect, an attenuation of an originally virulent culture can be obtained by passing the germs through animals with little or no susceptibility for diphtheria.  The so attenuated germs maintain their reduced virulence until they are placed again upon a more favorable soil, on which they may regain their full virulence.  These facts are of eminent importance for the preparation of diphtheria therapeutic serum, the principles of which may be stated in the following notes :

Preparation of diphtheria antitoxin.

We have mentioned that the quantity of antitoxin obtainable depends mainly upon the susceptibility of the animal.  Therefore one would have to select a species which warrants a good quality of serum by a high susceptibility for diphtheria poison.  The most useful objects for the purpose, especially in laboratory work are goats, which by Ehrlich and Wassermann are considered preferable to other animals, not only on account of the high susceptibility they possess, but also because they show a great resistance when sick from inoculations, and further afford an opportunity to use their milk which possesses considerable antitoxic value. Thus we have in these animals all the advantages combined which enable us to secure in regard to quality as to quantity the most efficient material.

The method of immunizing consists in créating a tolerance of the diphtheria poison by systematic inoculation with at first attenuated diphtheria cultures, later on more virulent cultures, and finally by direct application of the toxin. As the tolerance or resistance against the poison increases so does the immunizing power of the blood by production of antitoxin.

The toxin used for the inoculation is obtained in the following way: From a colony of diphtheria germs cultivated on a Petri dish a quantity is added to a quart of beef tea in a large flask. The latter remains three or four weeks in an incubator, during which time the germs saturate the fluid with their toxic products. By adding ½ per cent carbolic acid or ⅓ per cent trikresol the bacilli are then precipitated to the bottom of the flask and the liquid now contains the toxin in solution. From this solution the toxin can either be prepared as a dry substance, which is, however, very expensive, or it can be used while dissolved in the fluid after filtration through a Chamberland filter. The toxin is now ready for injection, the effect of which is local swelling, fever and the production of antitoxin.

The latter is found in the blood after each injection and increases in quantity with the number of injections. Valuation of the antitoxin. For the estimation of its efficacy it is necessary to tap the animal from time to time for a small quantity of blood with which experiments are made on others. The mode of valuation is based upon the observation of Behring and Kitasato that toxin and antitoxin neutralize each other when mixed in certain proportions in a test tube. Different from this is the method previously used, by separate injections of toxin and antitoxin, which however does not give as exact results on account of the varying absorbing ability of the animal. With administration of the former method an examination of the blood samples is made according to Ehrlich in the following way: Of a toxin of a standard strength 0.1 grm. representing the minimum fatal dose, for a guinea pig weighing 300 grms., 1 grm. is mixed with a number of different quantities of the blood to be examined, for instance, 0.4 grm., 0.3 grm., 0.2 grm., 0.1 grm., and with these four mixtures as many guinea pigs are inoculated. The result will be that corresponding with the amount of serum a different local and general effect takes place.

The animal which received the largest dose will show no reaction. The one that received 0.3 grm. of serum may suffer from an acute local inflammation, followed by necrosis, but otherwise be not much affected. The third will probably become quite sick, and the one which got the smallest dose of antitoxin will die within forty-eight hours from the effect of the same quantity of toxin as the other received. In this way we can determine which amount of blood of the immunized animal is sufficient to neutralize a certain quantity of toxin, and so the value of the serum can be established.

The absolute estimation, or an exact figure of its value, can be given, if one has a material of a standard strength with which to compare others. Behring and Ehrlich therefore prepared a serum of which 0.1 grm. is sufficient to neutralize 1 grm. toxin (ten times the fatal dose for a guinea pig of 300 grms.), and this they called "Normal Therapeutic Serum." Of such normal therapeutic serum 1 ccm. is termed the equivalent for one immunizing unit.

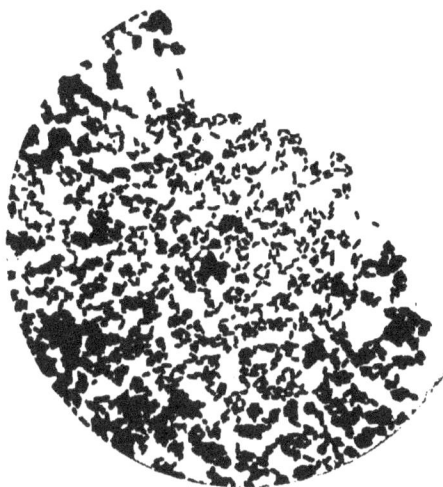

No. 6.  Diphtheria Bacilli.

Culture on agar-agar, twenty-four hours old ; stained in alkaline methylen blue, magnified 1,000 times.

Diphtheria bacilli is best cultivated on a substrate of agar, or blood serum.  Macroscopically they appear as a shiny, granulated mass ; microscopically they present short, irregular bacilli, frequently showing on one end the shape of an olive.

Accordingly a serum, of which 0.01 grm. is sufficient to neutralize ten times the fatal dose, would possess ten immunizing units or would be a ten fold normal serum; if 0.001 grm. has the same effect, 1 grm. of the serum would have 100 immunizing units, etc. Has the value of the serum once been established, the animal is tapped for a lárger quantity of blood, which, being received in sterilized vessels, is placed on ice until coagulation has been completed and the serum been separated as a clear, yellowish fluid. This fluid with 0.5 per cent carbolic acid now presents the material to be used for therapeutic purposes.

It is of special importance to remember, that in preparing the serum it is not our aim to accomplish a high degree of immunity in the animal, but to avoid this as much as possible because the longer the animal show its susceptibility, the better will be the serum it furnishes in relation to the increased production of antitoxin in its system. As the valuation of the serum is of vital importance for the practical administration, it seems justified to consider this subject somewhat more explicitly.

Although the principles of the above-mentioned method have been recognized as most practicable when the experiments with antitoxins were still in their earliest stage, special attention was given the subject by Behring and Knorr in their publication of ''Infection and Disinfection.'' Later the methods of valuation were repeatedly modified according to Behring-Wernicke's and Behring-Boer's investigations. After discovery of diphtheria antitoxin the method of mixing toxin with antitoxin was at first the one usually employed; for several reasons however it was temporarily abandoned; first, because the preparation of a toxin strong enough to effect death in very small quantity met at ˙ that time with great difficulties, and second, because it was believed that the degree of immunity obtained, would directly indicate the value of the antitoxin, produced in the animal, so that the strength of the serum could simply be determined by the amount of the inoculated poison. If this were correct it would be unnecessary to enter into the complicated examination of blood samples, by which toxin, antitoxin and animals are consumed. But the investigations of Behring and Wernicke concerning the relation of immunity to therapeutic results in diphtheria showed that a staple relation between the acquired immunity and the quantity of antitoxin does not exist. On the contrary, as was mentioned in the chapter of tetanus, Behring found that occasionally an animal may possess a large quantity of antitoxin and at the same time a higher susceptibility than before.

In face of these facts the method, based upon the idea that a certain relation between immunity and antitoxin existed, had to be abandoned. By later investigations, Behring and Boer attempted to solve another question, namely, how the effect of antitoxin in an animal, inoculated with *virulent cultures*, compares with the effect of an equal dose of antitoxin in an animal inoculated with the *ready toxin*.

Previous to this the efficacy of the normal serum was estimated by its power to save an animal after inoculation with ten times the fatal dose of a living culture, which was considered an analogical procedure to the infection of patients. On account of this analogy, it was believed that, after the difference of the effect in animals had been established, and so the necessary dose of antitoxin as a prophylactic and as a curative had been determined, the result would also apply to man. This conclusion, however, was found to be incorrect, and the result of many observations on persons showed, that the amount of antitoxin required to cure diphtheria infected patients, can only be estimated, aside from due consideration of the different weight, by comparative calculation of the quantity of *ready toxin* and not of the virulent cultures given to the animal.

How important this discovery was, is readily understood, if one compares the quantity of antitoxin required to save an animal after infection with *living cultures*, and that necessary to save it after inoculation with the *ready toxin*. A large series of experiments made in this respect, led to the result, that the dose of antitoxin necessary to neutralize ten times the minimum fatal dose of *toxin*, is fifty times larger than that for the equal amount of virulent *cultures*. In accordance herewith, Behring and Boer stated, that

1. 1 ccm. of their "Normal serum," injected separately from the poison, saves 5,000 grms. of animal from death, after inoculation with the tenfold fatal dose of a two days old diphtheria culture.

2. 1 ccm. of normal serum, injected separately from the poison, saves 100 grms. of animal from death, after inoculation with the tenfold fatal dose of toxin.

3. 0.1 ccm. of normal serum neutralizes in a test tube, at least ten times the fatal dose of toxin for a guinea pig of 300-400 grms. The accuracy of these figures necessitate of course, a steady toxicity of the poison and, knowing that the latter varies considerably, we would have to establish, what is the fatal dose or what is the strength of the toxin, to which the above figures apply. As the individual disposition represents a strong factor in the estimation of the toxicity, we can only calculate the fatal dose of taking the average from experiments on a larger number of animals. If, for instance, ten guinea pigs, weighing between 200 and 400

grms., or together 3,000 grms. were inoculate with 1
grm. diphtheria *toxin*, and three of them die, while the
rest after more or less serious affections, recover, we would
consider 1 grm. the minimum fatal dose for 3,000 grms.
living weight, or 0.1 grm. for one animal.

With reference to a diphtheria culture in beef tea the
fatal dose is considerable smaller, about 0.005 ccm., provided the culture was well developed by being kept in an
incubator for two or three days.

The quantity of toxin which represents the fatal dose
having been established, a complete valuation of a serum
would require four series of experiments, intending to determine :

1.  The immunizing value after an infection (inoculation with cultures).

2.  The curative value after an infection.

3.  The immunizing value after an intoxication.

4.  The curative value after an intoxication.

The results of such experiments are the figures mentioned above under 1 and 2. The most essential point of
these discoveries is the fact that

1.  The administration of antitoxin is perfectly harmless.

2.  It positively possesses prophylactic and curative
properties, if applied in the proper quantity.

Of what effect these properties of the serum would be
on men had of course to be investigated on patients. The
numerous observations in German and French hospitals by
noted medical authorities gave abundant evidence to cor-
robate the results, previously obtained on animals. In all
of these the correct estimation of the necessary dose was
instrumental not only in getting a definite judgment of the
merits of this treatment, but also in convincing those mem-
bers of the profession, whose doubts and prejudice against

No. 7.  DIPHTHERIA BACILLI.

Culture on blood serum, prepared as No. 6, magnified 1,000 times.

The short form presented in this specimen is due to their rapid multiplication.  Some of the germs are distinguished by a club formed shape, which is considered characteristic to this species.

this therapy rendered them unable to comment upon the value of the so minutely conducted experiments. Thousands of animals have been sacrificed in order to arrive at certain conclusions about the utility of this therapy and having passed the experimental stage it was given a trial even by the most conservative scientists. The clinical experience of many observers extending over several years not only unanimously agrees upon the high practical value of the new doctrine but is in itself a contribution to our knowledge as to the dosage and the therapeutic limits of blood serum therapy.

We now will direct our attention to the establishment of the correct dosage and to certain conditions to be considered in the administration of this therapy. As explained above, a much larger dose of therapeutic serum is necessary to neutralize the effect of an intoxication than that of an infection. This appears quite natural as the serum acts in both cases against the toxin present, which after an infection is but originating and therefore easier neutralized than the corresponding dose of ready toxine. For the same reason it is evident that it requires a larger dose of antitoxin to cure a case of several days' standing than to check an infection in its incipiency, because in the former the bacilli have produced considerable more toxin. How much more antitoxin is required in an advanced case of diphtheria cannot be exactly stated and depends mainly on the gravity of the case, but it is certain that the effect is proportional to the dose, as noticed on animals, which received more antitoxin than necessary to neutralize a previous diphtheria infection. For instance, a guinea pig, if infected with ten times its fatal dose, would be saved by a quantity of antitoxin in the proportion of 1 to 5,000 of its weight, but not without showing signs of severe illness. The latter would be materially lessened by a dose of the

*(margin note: Dosage of antitoxin.)*

proportion of 1 to 2,000 of animal weight and may not set in at all after a dose of 1 to 500.

The experiments made on children with blood serum showed that at least 500 immunizing units are necessary to effect a cure even in light cases of diphtheria, that is 50 ccm. if the material is a tenfold normal 'serum, or 10 ccm. if it is fifty times as effective. In more serious or advanced cases 1,000 to 1,500 immunizing units would be required to save the child and as there are many factors which may be apt to aggravate a case, it will always be a matter of personal judgment how to class the respective case.

An item also worthy of consideration is the fact that several cases being apparently under the same condition, take an entirely different course. In the same family, for instance, one child may present only the common picture of angina and recover in a few days. Another child may have laryngeal membranes which necessitate tracheotomy or intubation. A third one may show from the beginning a serious affection of the tonsils, throat and nose with obstructions by thick membranes and suppuration from the effect of which it could not be saved even by early tracheotomy. These cases, though having originated from the same cause, the infection with diphtheria germs, would require very different doses of antitoxin. In order to determine the necessary dose it must be remembered that the latter should be proportional to the toxin present which as is well known increases rapidly from day to day after an infection.

A curative effect can therefore with certainty be expected only if the organism has not too long been preoccupied by the toxin, that is, in an early stage, or in a milder form of the disease. When the diphtheric process already affects the bronchi and lungs, so that even tracheotomy is of no avail, the application of serum will hardly do better.

The prognosis is always doubtful if the toxin has circulated for a longer period, say three to four days, because their paralyzing effect upon the nerves and ganglia of the heart can no more be eliminated. The most important factors of aggravation of a case is the complication by a simultaneous infection with other germs, especially streptococci, and it is in such cases of mixed infection that we find the most virulent diphtheria bacilli.

In order to intelligently comprehend the character of these cases and to express an opinion as to the utility of antitoxin it seems necessary first, to understand the position and effect of the nondiphtheric microbes. According to the observations of Barbier, Schrader, Roux and Yersin the infection with diphtheria bacilli present a more serious character when mixed with other germs, especially streptococci. The latter penetrate the mucous membrane of the tonsils and are carried to the lymphatic glands which, affected by the bacterial products become œdematous and finally turn into suppuration.

Aside from these complications the diphtheria bacilli become more virulent in the presence of streptococci. Starting from this fact Funk tried to ascertain whether the diphtheria bacilli really produce a stronger toxin or if the organism is rendered more susceptible for the poison by said complications. He inoculated twenty-two guinea pigs with a certain multiple of the fatal dose of diphtheria culture, to which in eleven animals a culture of streptococci was added. Twenty-four hours before the inoculation all animals had received a sufficient quantity of antitoxin to neutralize the diphtheria infection. The result was, that the animals with mixed infection died, while the others recovered. If instead of a living culture the ready toxin was applied the streptococci had no aggravating effect. It was proved by these experiments that streptococci render the diphtheria bacilli more

Antitoxin in mixed infect

virulent. Knowing this fact, one should expect that an accordingly larger dose of antitoxin would balance the
increased toxicity, and to a certain extent this is really the
case.

But as mentioned above the streptococci produce other
symptoms which are apt to decrease the vitality and resistance of the patient. Against these complications nothing
can be expected from the antitoxin. It may, however, be
possible to check a mixed infection by application of two
different antitoxins, one acting against the diphtheria toxin, the other intended to neutralize the effect of the products of streptococci. Being at present engaged in these
experiments we hope to come soon to a conclusion as to
their practical value.

From the facts above mentioned we arrive at the following conclusions in regard to serum therapy : 1. Every
case should be treated in its incipiency locally with such
antiseptics as seem fit to suppress a mixed infection.
2. The antitoxin will be of so much better effect even in
cases of mixed infection the sooner a proper dose of antitoxin is applied.

Local treat-
ment.
The local treatment with antiseptics is not to be neg-
lected for another reason. The germs developing on the
tonsils extend their effect not only to the case in question,
but may be instrumental for the infection of others. Even
convalescents carry the virulent germs sometimes for a long
period after all local symptoms have disappeared. In some
instances diphtheria bacilli were found as long as eight
weeks after recovery, a fact which deserves special attention in reference to school hygiene. In other cases an
apparent recovery is followed by secondary affections of
the kidneys, peripheric nerves, etc., also due to diphtheria
intoxication, and in face of such complications it may seem
that the virtues of antitoxin were but very limited. Con-

sidering, however, that if the production of the toxin is
neutralized in an early stage, such serious results are not
very liable to occur, we come again to the conclusion that
everything depends upon the early application of the antitoxin as well as an appropriate local treatment.  In regard
to secondary affections the experiments of P. Meyer have
shown that diphtheria paralysis is due to degeneration of
nerve fibers, caused by the toxin.  Is the latter effectually
checked by antitoxin, the patient will have a good chance
to escape such complication.

Knowing the various factors which are of importance
regarding the efficacy of serum therapy we will be able to
consider more intelligently the results so far obtained.
According to Roux and Martin's observations in Paris the
effect of antitoxin was most favorable in cases in which
tracheotomy was performed.  The mortality of these cases
averaged during the last five years about 85 per cent, and
was reduced by serum therapy to less than 47 per cent.
At the same time many others which previously would have
had to be operated were cured by this new therapeutic
agent without surgical aid.

Even better results than these are shown by statistics Statistic.
from the hospitals of Berlin. In most of the cases the diagnosis was made, based on bacteriological examinations and
such cases which started with diphtheria symptoms, but
later proved to be scarlet, measles, etc., were not considered
in the following figures : Of 233 cases reported by Kossel
179 recovered, which is equal to 77 per cent.  In these are
included 72 children with tracheotomy, of whom 41 recovered, representing 57 per cent against but 25 per cent in
former years.

To show how the results of serum therapy differs according to the age of the patient and stage of the disease,
we add the inclosed tabulæ.  From tabula 1 it is evident

that the high percentage of recoveries after tracheotomy is chiefly due to the good results obtained in children under two and five years, while without operation the older ones show better results than the others.

More interesting than this is tabula 2, which represents the same 233 cases divided into groups according to the day when the treatment was commenced. The results obtained in the first two days, speak for themselves, and need no further comment. The statement relating to the advanced cases is all the more remarkable as the determination when a case began, depended mainly upon the dates given by relatives, and for this reason it seems just to presume, that in many cases the disease had started sooner than observed by them.

The results of serum therapy, as given in tabula 2, are in harmony with the effect seen in experiments. The longer the toxin had circulated in the organism, and the more complicated the case was, either on account of obstructions in the respiratory organs or on account of the virulence of the infection, the less effect had the antitoxin. The antitoxin used in the cases reported in the above tabulæ, was fifty to sixty times as strong as Behring's normal serum, so that 1 ccm. contained fifty to sixty immunizing units.

Effect to local symptoms.

The dose administered, varied according to the case between 160 to 800 units. The injections were usually made on the lateral part of the chest. The fluid was absorbed within one to two hours, producing more or less pain, which gradually disappeared in course of twenty-four hours. An immediate reaction regarding the temperature has in no case been observed, but in many instances a skin eruption like urticaria appeared, several days and as late as two weeks after the injection. This has more frequently been noticed when the serum of

# RESULT OF ANTITOXIN IN DIPTHERIA.

## TABULA I.

| Age. | Number Treated. | Recovery. | Death. | Percentage of Recovery. | With Tracheotomy. Number. | With Tracheotomy. Recovery. | With Tracheotomy. Death. | Percentage of Recovery. |
|---|---|---|---|---|---|---|---|---|
| 0-1 | 1 | ..... | 1 | ..... .... | 1 | ..... | 1 | 0 |
| 1-2 | 16 | 8 | 8 | 50 | 8 | 2 | 6 | 25 |
| 2-3 | 35 | 27 | 8 | 77 | 14 | 8 | 6 | 57 |
| 3 4 | 40 | 30 | 10 | 75 | 17 | 13 | 4 | 77 |
| 4-5 | 34 | 24 | 10 | 70 | 11 | 8 | 3 | 73 |
| 5 6 | 23 | 19 | 4 | 83 | 5 | 2 | 3 | 40 |
| 6-7 | 16 | 10 | 6 | 62.5 | 8 | 4 | 4 | 50 |
| 7-8 | 21 | 16 | 5 | 76 | 4 | 0 | 4 | 0 |
| 8-9 | 19 | 17 | 2 | 89.5 | 3 | 3 | 0 | 100 |
| 9 10 | 12 | 12 | 0 | 100 | 1 | 1 | ...... | 100 |
| 10-11 | 6 | 6 | 0 | 100 | ...... | ...... | ...... | ...... |
| 11-12 | 7 | 7 | 0 | 100 | ...... | ...... | ...... | ...... |
| 12-13 | ...... | ...... | ...... | ...... | ...... | ...... | ...... | ...... |
| 13-14 | 2 | 2 | ...... | 100 | ...... | ...... | ...... | ...... |
| over 14 | 1 | 1 | ...... | 100 | ...... | ...... | ...... | ...... |
| | 233 | 179 | 54 | | 72 | 41 | 31 | |

## TABULA II.*

| Duration in Days. | Number Treated. | Recovery. | Death. | Percentage of Recovery. |
|---|---|---|---|---|
| I. | 7 | 7 | 0 | 100 |
| II. | 71 ( 9 ) | 69 ( 7 ) | 2 ( 2 ) | 97 |
| III. | 30 ( 7 ) | 26 ( 6 ) | 4 ( 1 ) | 87 |
| IV. | 39 (14 ) | 30 (10 ) | 9 ( 4 ) | 77 |
| V. | 25 (11 ) | 15 ( 5 ) | 10 ( 6 ) | 60 |
| VI. | 17 ( 7 ) | 9 ( 2 ) | 8 ( 5 ) | 47 |
| VII. | 41 (23 ) | 21 (10 ) | 20 (13 ) | 51 |
| VIII. | 3 ( 1 ) | 2 ( 1 ) | 1 | .... .... |
| | 233 (72 ) | 179 (41 ) | 54 (31 ) | 77 |

*The figures in parentheses refer to cases in which tracheotomy was performed.

sheep or dogs, than if that of goats or horses had been used. It proved to be in all cases, a very harmless occurrence, which had no relation to the efficacy of the serum.

With reference to the local symptoms, a direct change after the injection of serum does not take place. The white membranes were frequently found larger on the next day, than they were before, but notwithstanding this apparent progress, the disease had not advanced. Especially the larynx, became in no case involved after the application of antitoxin, if it had not been affected previously, and even many of those in which laryngeal symptoms were present, recovered without tracheotomy. More visible was the effect in regard to the general condition of the patients, characterized by pulse and temperature. In fresh cases it was observed that both showed a critical decrease after large doses of serum. The conclusion regarding the efficacy of antitoxin, judged by the condition of pulse and temperature alone, is however sometimes incorrect ; first, because a high temperature frequently remains, after the child has entered the convalescent stage, and second, because the fever may be due to a complication, and not to the toxin of diphtheria. Pulse and temperature should therefore not be expected to decrease rapidly, even if the antitoxine has its full specific effect. A better sign by which to judge the latter, is the subjective condition of the patient. It is remarkable how in cases of a most severe form, evident signs of distress vanished within twenty-four hours and the children whose physical and mental condition was previously very low, showed strength and good humor. In such cases it was naturally surprising to the physician in charge " to see children, who were in doubtful condition when admitted,

Effect to general symptoms.

playing with their toys or eating with apparently good appetite the following day." Regarding those cases which resulted fatally, notwithstanding the application of antitoxin, it seems that at least in some of them, the dose was not adequate to the gravity existing, and we may therefore confidently expect still better results than so far obtained, after experience has enabled us to state with more accuracy the quantity required in the different forms which may come before us.

Rules for administration.

The principal conclusions to which the mentioned results have led in serum therapy are:

1. The chance to save a patient is the better the sooner antitoxin is applied.

2. The seriousness of a diphtheria infection is frequently underestimated, and for this reason it seems advisable to apply rather large than insufficient doses, for light cases 200 to 300 immunizing units, for severe ones, especially after tracheotomy, 500 to 800 units.

3. The application of the serum has to be repeated, according to the gravity of the case, on the next day or even on the same day until a quantity of 600 to 1,000 or more units has been injected.

4. The administration of serum of high potence is preferable to the corresponding amount of a weaker preparation.

According to this, Behring, whose method we have adopted, recommends a serum of sixty immunizing units, which is supplied in bottles of three different sizes, each being one dose. No. 1 representing 600 units to be applied only on the first or second day of the disease, No. 2 containing 1,000 units to be used in more serious cases on the first or second day or in less severe cases of longer standing. No. 3 possesses 1,500 to 1,600 units and is meant for adults or very severe cases in children.

In case the serum is not intended as a curative but as a prophylactic, the strength and quantity required is considerably smaller. In the average 100 to 200 immunizing units are sufficient to protect a healthy child against the infection. How long such protection lasts has not been fully determined, and Kossel advises a reinoculation after two to three weeks if the child is still exposed to infection. Such small doses, however, are insufficient if the child is in the period of incubation, which would put it under the class of an early stage. It sometimes may happen that children are infected, but for lack of visible symptoms do not get an adequate dose, in which case the failure is to be attributed to the fault of judgment and not to the inefficiency of the serum. The question regarding the dose required for prophylactic purposes has been discussed during the last few months with considerable interest. In a recent publication Behring shows how he came to use the quantity now recommended for immunization. Starting with one to five immunizing units in districts where the schools had to be closed on account of an epidemy, he soon was convinced of the insufficiency of such small doses. Better results were obtained with fifteen units, but in the stage of incubation even this dose proved inadequate. A further increase to sixty units gave uniformly good results, and the fact that of 10,000 persons inoculated with this amount a few took sick, would not justify us in assuming that a larger quantity is required as the normal dose for immunization.

But another point has to be considered in regard to this question. As the immunity gradually decreases according to the elimination of the antitoxin, if no additional injections are given, it is evident that its duration also depends upon the quantity first administered. On the other hand, one can maintain an immunity for a longer period by repeated application of smaller doses in certain intervals. For this reason a dose not over 150 immunizing units has been recommended by Behring for the purpose of immunizing healthy persons. If the serum in this quantity or any other, which future experience might determine, comes up to our expectation regarding its immunizing power, this method will prove to be the strongest factor to prevent the disease. A most essential auxiliary will always be an early diagnosis and the immediate application of an appropriate local treatment.

Production of antitoxin.

To assist in the complete understanding of serum therapy a few more words may be said about the present theories regarding the production of antitoxin, and its effect upon the toxin in the organism. What has been said in regard thereto in the chapter of tetanus applies as well to diphtheria. Comparing the opinion first expressed by Behring and Kitasato, that the toxin and antitoxin have a direct effect upon each other because of their being neutralized when mixed in a test tube in certain proportions, and the opinion of Buchner and others that the efficacy of antitoxin must be attributed to an action of the living cells, we will find that either of these opinions is correct to a certain extent. The cells play undoubtedly an important part in effecting the results observed, but at the same time a direct antagonistic action must exist between toxin and antitoxin. This is corroborated by the test which shows how an inoculation with a mixture of toxin and antitoxin compares with the separate application of the two substances. In the former case two parts of the antitoxin mixed with the toxin proved sufficient to neutralize the toxin, while five times as much was necessary to save the animal if both substances were injected on different places.

Action of cells in immunization.

The most evident proof for the action of the cells is the effect of a mixture of toxin and antitoxin upon different

species of animals. If these substances were neutralized
in vitreo, the toxins could have no effect upon any animal
injected with such mixture. It has been found, however,
that if such mixture was ineffective for instance upon
mice, it still acted as a poison upon guinea pigs on account
of their higher susceptibility. Therefore, one cannot deny
that the cells are instrumental in producing the antitoxin.
If the production of the antitoxin would belong to the
blood, the serum of an animal could not very well maintain any immunizing power after a quantity as large as the
animal's capacity had been drawn. The serum, however,
retains not only its previous immunizing value, but shows
an increase of the latter after every additional application
of poison. The latter, therefore, seems to act as a stimu-
lant to the antitoxin producing cells.

The idea that antitoxin is a cell product is further
substantiated by F. Klemperer's observation that in immunized chickens the yolk but not the white of the egg has
antitoxic properties.

With the above explanations we have outlined the
principal points of serum therapy as far as they are of
interest to the practitioner, and in conclusion wish to
make a few remarks which may be of practical value.

With reference to the preparation of the serum it
should be borne in mind that in order to obtain a serum
of high value, the animal which is to furnish the serum
must be inoculated with either highly virulent cultures or
a strong toxin. If the former is used it has to be frequently tested for the degree of its virulence, which is
known to change from very slight causes. We have observed that a temporary placing of the culture in a re-
frigerator decreases its virulence to a considerable extent,
and that even a difference in the temperature of the laboratory sometimes leads to very different results concerning the efficacy of a culture.

Of equal importance is the proper composition of the nutrient substrate, and its reaction which even after being properly prepared is subject to alterations caused by the development of the germs. All these difficulties are avoided if ready toxin of a tested strength is used for the inoculations because its virulence is not altered by cold or other atmospheric influences. Furthermore the serum of the animal herewith intoxicated must be tested as to its value on a series of animals after each injection, and also each time it has to furnish serum for therapeutic purposes. This is the most important part in the preparation of the antoxin, because the clinical results depend upon a correct estimation of the material put on the market.

A source which cannot be trusted in this respect should, therefore, be boycotted by the profession. The result of such tests should be stated on the label of every bottle or vial (cf. sample), giving the number of immunizing units per ccm., so that the practitioner is able to tell what dose to inject without depending on the directions, which cannot apply to every case uniformly. It seems advisable to furnish at least two different sizes of the material, one for prophylactic and one for curative purposes, each representing one dose in order to avoid a possible infection of the serum by repeated opening of the bottle. With reference to the effectual suppression of diphtheria epidemies, the sanitary officers should coöperate with the practitioner by publishing and distributing, especially among the laboring classes, such instructions as would inform them of the possibility to prevent the disease, and to call their attention to the necessary therapeutic measures. At the same time these instructions would help to overcome the prejudice, so common among the lower classes, against any operative procedure.

We advise a distribution of a circular with the following information :

1. Diphtheria is an infectious disease most dangerous to children, which can be cured by early attendance.

2. The usual symptoms are first a swelling of the tonsils, commonly known as sore throat. Soon after the inflammation of the tonsils white spots appear on the latter, causing destruction of the tissue and frequently obstructing the air passages.

3. These white spots or membranes contain germs which, when expectorated, may infect other persons with the disease.

4. With the inflammation of the tonsils fever and general discomfort set in.

5. From the tonsils the disease frequently extends to the nose, from which a suppurative fluid is discharged, and to the throat and lungs, causing hoarseness and difficulty in breathing.

6. The further effect of the disease is inflammation of the lungs, kidneys, and a general poisoning of the system, which may result in a sudden paralysis of the heart.

A cure of this disease can be effected almost with certainty when the treatment is applied in the incipient stage —that is, the first, second or third day after the earliest symptoms appeared. In advanced cases at least a diminution of the gravity and the danger to infect others may be effected by the same treatment. It is, therefore, the duty of the parents or guardians of children suffering of diphtheria to apply at once after discovery of the first symptoms to the proper authority for attendance.

1. The treatment consists in hypodermic injections of a remedy called "antitoxin," which has the power to nullify the poisonous effect of the disease. This remedy, as well as its application, are perfectly harmless.

2. With the same remedy in smaller dose other mem-
bers of the family can be protected against the disease.

3. The sick child should, if possible, be isolated from
others.

4. All clothes and bed linen used in the sick room
must not be moved from the room or used before being
sterilized according to the directions of a physician.

5. The person who attends to the child should frequently wash his hands and face with a disinfecting solution as prescribed by the attending physician.

6. All dishes and table supplies used by the child
should be placed in boiling water immediately after use.

We are confident that such measure will prove of valuable assistance in securing satisfactory results from the use
of antitoxin, as the coöperation of the laity is necessary to
obtain as good effects as were accomplished in foreign hospitals, in which the largest percentage of recovery occurred
in those children who were inoculated in the earliest stage
of the disease. The enlightenment of the public on this
topic will undoubtedly assure still better statistics regarding the success of blood serum therapy.

### BIBLIOGRAPHY.

Archiv. f. Klin. Chirurgie 1888-1892.
Archiv. of Pedriatrics Bd. X., 1893.
Annales de l' instit., Pasteur, 1890-94.
Behring, Blutserum Therapie, Th. 1 and 2.
Behring and Knorr, Infection and Disinfection.
Behring, gesammelte Abhandlungen.
Berliner Klin. Woch., 1889-1894.
Brieger, Ptomaine Bd., 1, 2. 3.
British Med. Journal, 1891-1894.
Centralblatt f. Bacteriologie, 1887-1894.
Deutsche Med. Woch., 1889-1894.
Emmerich and Terboi, Natur der Schutz undHeilsubst des Blutes.
F" "ænkel, Bacterienkunde.
Giorna dell R. Accad. d. Med. di. Tarr. 1884.
La Riforma med., 1891-1894.
Nicalaier Inaug. dissert. Goettingen, 1885.
Philadelphia Med. News, 1893.
Zeitschr f. Hygiene und Infectionskr., 1889-1891.

## SAMPLES OF LABELS.

---

NAME OF FIRM.

5 CCM.

# DIPHTHERIA-ANTITOXIN
CONTAINING
## 30 IMMUNIZING UNITS PER 1 CCM.

TO BE USED HYPODERMICALLY AS ONE DOSE AS A
PROPHYLACTIC AGAINST DIPHTHERIA.

---

NAME OF FIRM.

10 CCM.

# DIPHTHERIA-ANTITOXIN
CONTAINING
## 50 IMMUNIZING UNITS PER 1 CCM.

TO BE USED HYPODERMICALLY AS ONE DOSE IN
INCIPIENT CASE OF DIPHTHERIA.

---

NAME OF FIRM.

20 CCM.

# DIPHTHERIA-ANTITOXIN
CONTAINING
## 50 IMMUNIZING UNITS PER 1 CCM.

TO BE USED HYPODERMICALLY AS ONE DOSE IN
ADVANCED CASES OF DIPHTHERIA.

www.ingramcontent.com/pod-product-compliance
Lightning Source LLC
Chambersburg PA
CBHW021954190326
41519CB00009B/1251